I0446890

Hydrogen Generator Instructions

These plans can be used to build a hydrogen assist generator that gets added to your vehicle's engine to increase your gas mileage. Setup and used correctly, this system will increase your gas mileage up yo 30% or more!

These plans are complete; from building the reaction chamber to installing on your vehicle and how to use it.

Step-by-step picture and description build of the reaction chamber are included with these plans. Very easy to build with no mechanical knowledge needed.

REQUIREMENTS:

Read the entire instructions before starting anything.

Anything in red is indicating hazards or danger, pay special attention to this. BE CAREFUL, DANGEROUS CHEMICALS AND GASES. **I recommend putting this system onto a second vehicle until you have perfected it. This system is considered purely experimental and there 's more work that needs done to reach perfection.

SAFETY ISSUES:

Sodium Hydroxide is the active chemical that works with aluminum to release hydrogen from the water. Hydrogen gas will be created with this system; hydrogen gas is explosive and will burn. (that's how an engine uses it) Install this system outside as excess hydrogen could be generatedAND COULD CREATE AN EXPLOSION HAZARD IF DONE INDOORS. Do not attempt to do this unless you understand all the risks. I do not accept responsibility for any loss of life, limb, injury or property because of the use of this system, this system is considered experimental. Hydrogen burns with a

Hydroxide (LYE) is highly corrosive and will cause chemical burns if you get it on your skin, it can cause blindness if you get it in your eyes. READ THE MSDS SHEETS

Wear safety goggles to protect your eyes. Wear gloves to protect your skin.

BUILDING THE REACTION CHAMBER:

Make a list of the things that you will need.

TOOLS:

1. Knife to cut hose
2. Drill and bit (13/32) to drill a hole into the reaction chamber and air cleaner housing
3. Saw that will cut 4" PVC pipe
4. 1/4 - 18 tap

Materials:

1. Sodium Hydroxide (LYE)
2. Silicone RTV to seal around fittings (use automotive type)
3. Vaseline or grease to seal threads on screw cap
4. Small - medium sized plastic bottle
5. Zip ties to secure reaction chamber and backfire safety device
6. 15 to 20 inches of 4" SCH 40 PVC pipe; you will cut this to your length (longer is better)
7. (1) SCH 40 PVC solid end cap
8. (1) SCH 40 PVC end cap with threaded clean out
9. PVC prep solvent and glue
10. Scrap aluminum

YOU WILL ALSO NEED...

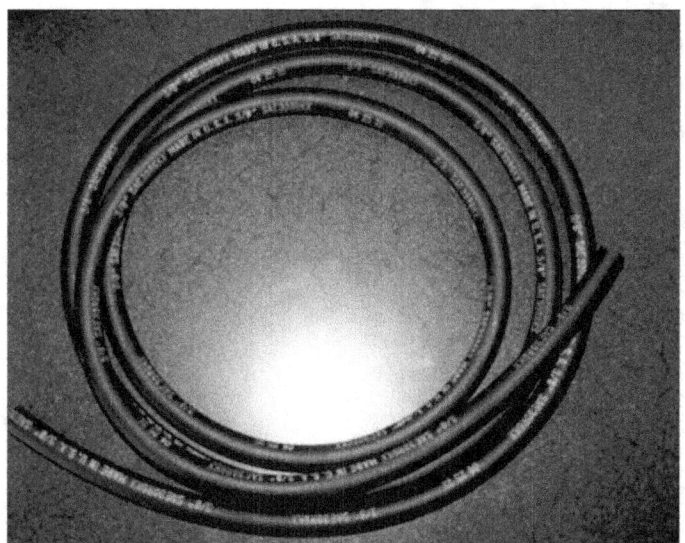

FUEL LINE (10 FEET) --- (You can get this at your local auto parts store)

VACUUM CHECK VALVES (2) --- (

HOSE CLAMPS FOR 3/8" HOSE

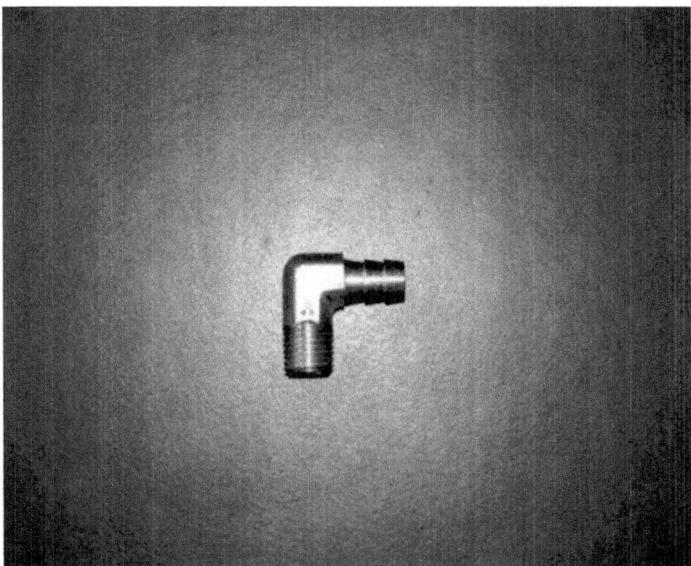

1/4" THREAD BY 3/8" HOSE END (Straight fitting--1; 90° fitting-- 2)

Start by measuring the clearance that you have to mount the reaction chamber. (leave room for the end caps and fittings) Cut the PVC pipe to the length that you need.

Use PVC prep solvent and glue (follow your product directions) to attach the bottom solid end cap, then repeat for the top threaded cap. After glue is cured, drill a hole on the side of the reaction chamber, at the top. (see picture link below) Drill a hole into the top center of the screw cap.

Install a 90° fitting into the hole that you drilled in the screw cap. (this will be a tight fit) Seal around the fitting with the silicone RTV. Install a straight fitting into the hole that you drilled in the reaction chamber. (this will be a tight fit) Seal around the fitting with RTV. Use the RTV and use around the seal where the PVC pipe and end caps join together.

THE REACTION CHAMBER MUST NOT HAVE ANY LEAKS!!

BUILD UP OF THE REACTION CHAMBER PICTURES ARE AT THE END OF THESE PLANS:

INSTALLING THE SYSTEM:

DO NOT MOUNT THE REACTION CHAMBER INSIDE PASSENGER COMPARTMENT, EXPLOSION HAZARD. Mount the reaction chamber under the engine bay away from the exhaust pipe. DO NOT MOUNT THE REACTION CHAMBER WITHIN 12 INCHES OF EXHAUST MANIFOLD SYSTEM, TOO MUCH HEAT!!

1. Zip tie the reaction chamber to a secure point on the vehicle. Try to mount the reaction chamber at a slight angle; this will minimize the splash when you fill it.
2. Cut about a foot of hose for the fitting on top of the reaction chamber. Clamp one side of the hose to the fitting, install and clamp one of the check valves to the other end. When installing the check valve, the arrow MUST be facing towards the reaction chamber. If there is no arrow, blow into each end of the check valve.   When you can blow threw the check valve, that is the direction the air will flow. When you put the cap on, put this hose somewhere that it is not against the reaction chamber. You do not have to secure this hose with zip ties. From the remaining hose, clamp one end of the hose to the fitting from the side of the reaction chamber. For the next step, you will have to cut the hose to your length.
3. You will have to install a check valve from the reaction chamber to the water lock. Install with the arrow pointed towards the water lock tank. (see illustration below)
4. The next step is to make a backfire safety device. The following illustration will show you how to make this. Locate this away from hot or moving parts.   The purpose of a backfire safety device is to filter the unwanted gases coming from the reaction chamber, it also acts as a safety devise incase of an engine backfire. Once you make the safety device, zip tie it to a secure location.

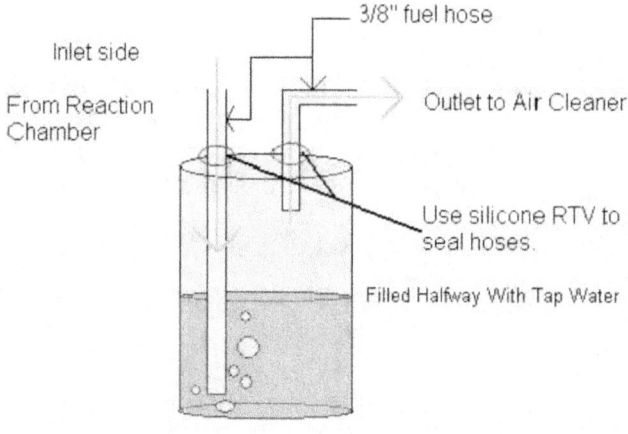

Inlet side

From Reaction
Chamber

3/8" fuel hose

Outlet to Air Cleaner

Use silicone RTV to
seal hoses.

Filled Halfway With Tap Water

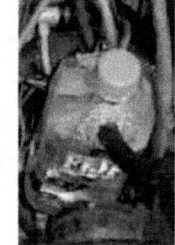

Backfire safety device
Made from plastic bottle

Gas Flow And Assembly

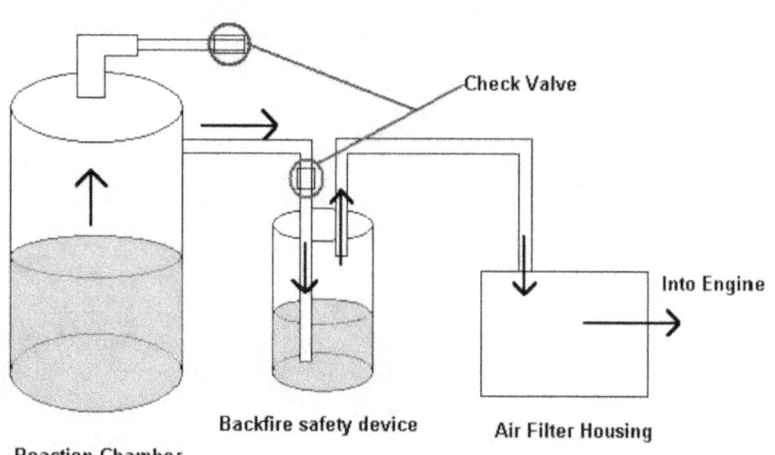

Check Valve

Into Engine

Reaction Chamber

Backfire safety device

Air Filter Housing

IMPLEMENTING THE DEVICE   :

1. With a 13/32 drill bit, drill a hole into your air filter housing.   You may want to remove
   your air filter housing before drilling to keep debris out of your engine.   Install fitting
   into your air filter housing and seal it with RTV. (this will be a tight fit) Install and
   clamp hose onto the fitting.

2. Fill the reaction chamber with one to two quarts of water. (depending on how big the reaction chamber is. Never fill the reaction chamber more than halfway .) lock tank half way with water. Distilled water will work better in the reaction chamber.

3. Now it's time to take your car outside if it isn't already. Put your safety goggles and rubber gloves on at this point. Use 10 tablespoonfuls of lye to 1 quart of water. Mix the lye with the water in the reaction chamber. Once you mix the lye with the water, let it settle down after the initial mix up, 5 to 10 minutes. Whatever aluminum you have can now be added. Be careful to not splash. Most aluminum cans have paint or coatings that turn into gunk that floats around and slows down the process. Unpainted aluminum is best. After adding aluminum you will start to see it bubble and hiss, this is the hydrogen being generated.

4. Screw the end cap into the reaction chamber. (the threads should be coated with Vaseline or grease to seal) You should be ready to go for your drive at this point.

TESTING STAGES AND ISSUES:

All experiments have their growing pains and some things will work better than others. You will find what mixtures work well, how many cans / grams of aluminum per 100 miles of travel for your engine.

You will have to add water and aluminum to continue reaction, but no additional sodium hydroxide (lye) is needed.

Another problem is the reaction process will continue when you turn your car off until all the aluminum or water is used up. This is the main reason this process is done outdoors so extra hydrogen is ventilated away into the atmosphere.

Hydrogen Generator Reaction Chamber Build

Hydrogen Generator Reaction Chamber Build

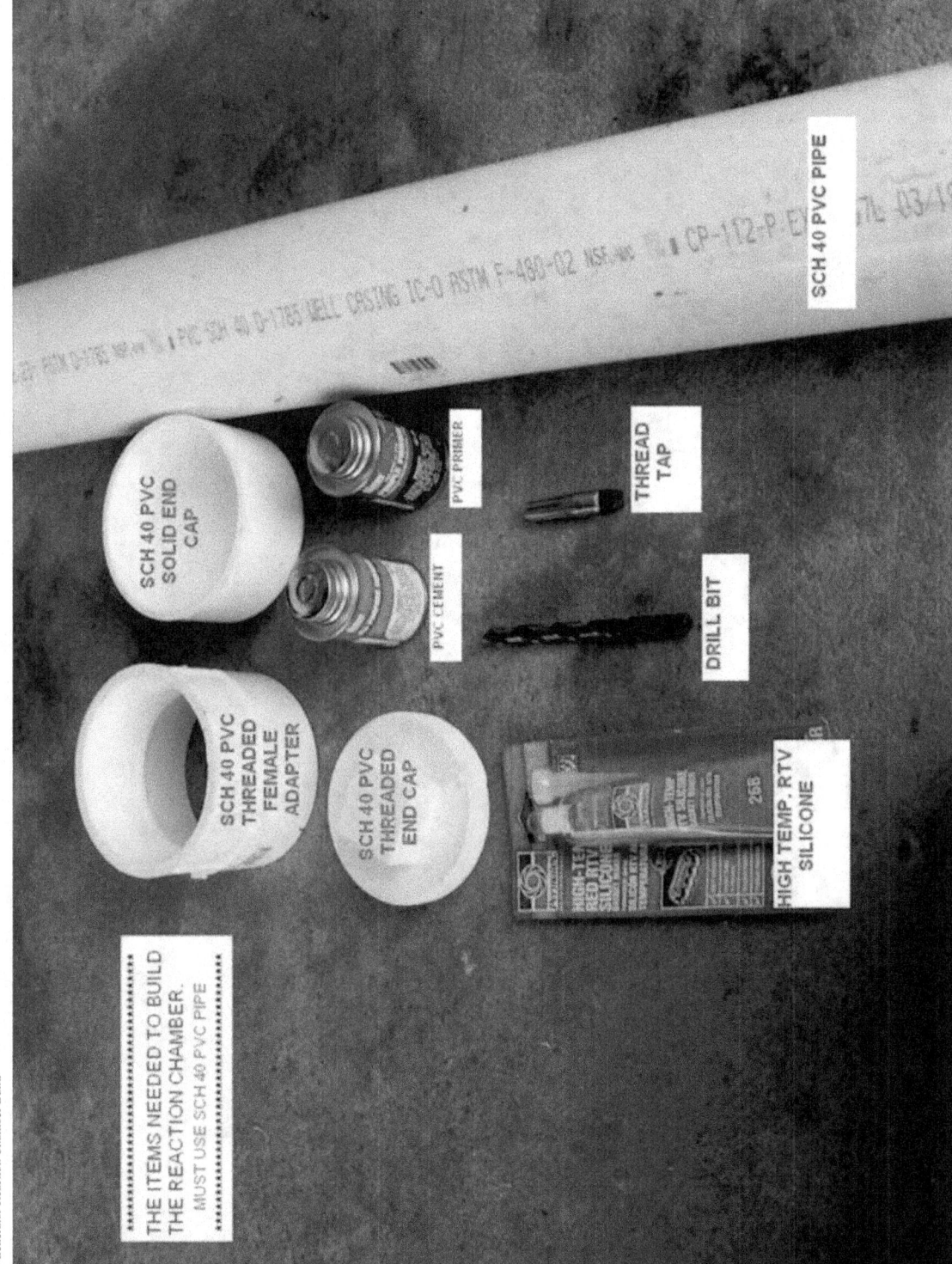

SCH 40 PVC PIPE

SCH 40 PVC
SOLID END
CAP

PVC PRIMER

THREAD
TAP

PVC CEMENT

DRILL BIT

SCH 40 PVC
THREADED
FEMALE
ADAPTER

SCH 40 PVC
THREADED
END CAP

HIGH TEMP. RTV
SILICONE

THE ITEMS NEEDED TO BUILD
THE REACTION CHAMBER.
MUST USE SCH 40 PVC PIPE

Once you cut the PVC pipe to the length that you want, follow the directions on the primer can and coat the one end of the PVC pipe as shown. Then follow the directions on the PVC cement can and install the threaded female adapter.

Threaded female adapter installed

Turn the PVC pipe around and repeat with the solid end cap

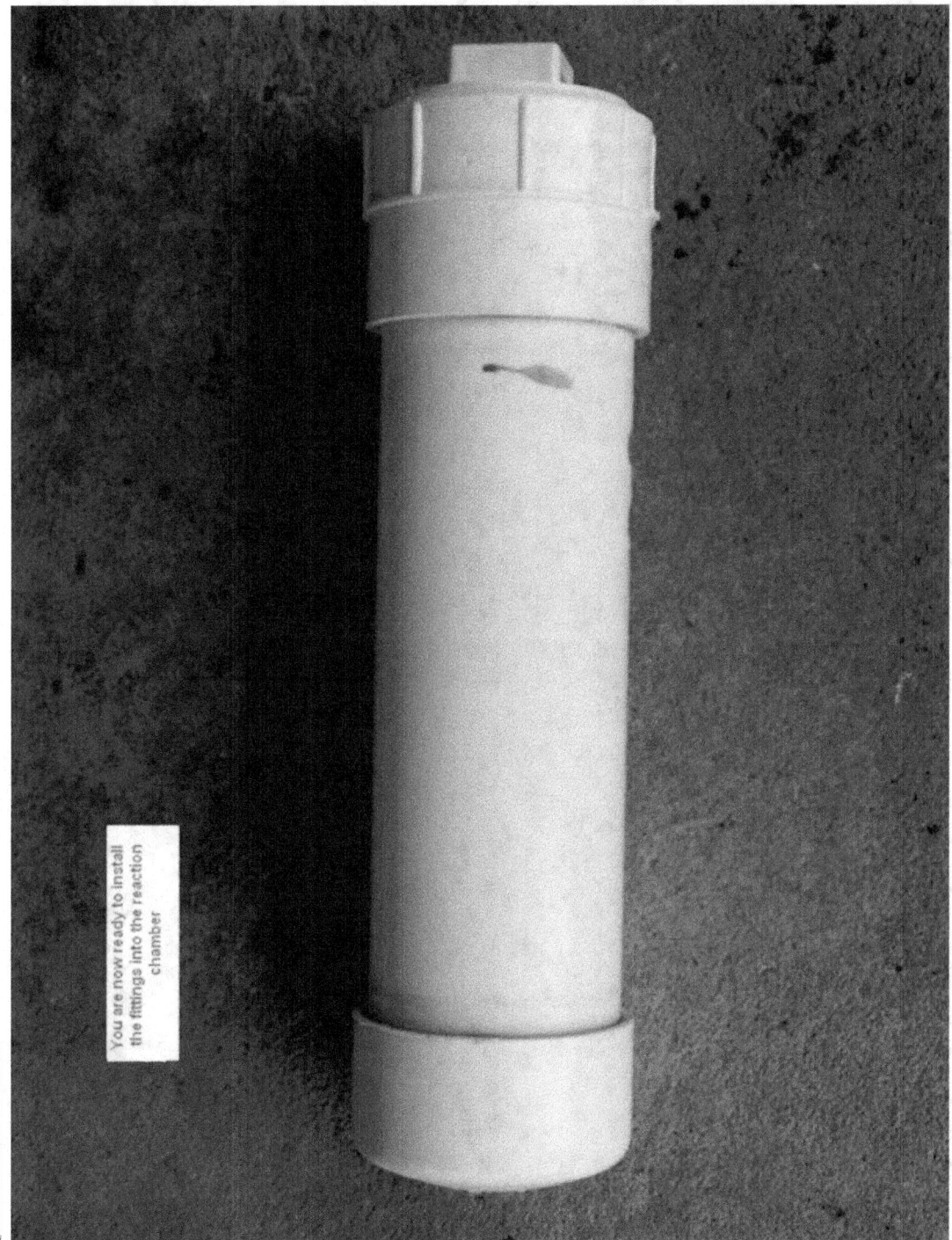

You are now ready to install the fittings into the reaction chamber

Using a 13/32 drill bit, drill a hole in the lower half of the threaded adapter end. Make sure that you drill into the adapter end and the PVC pipe.

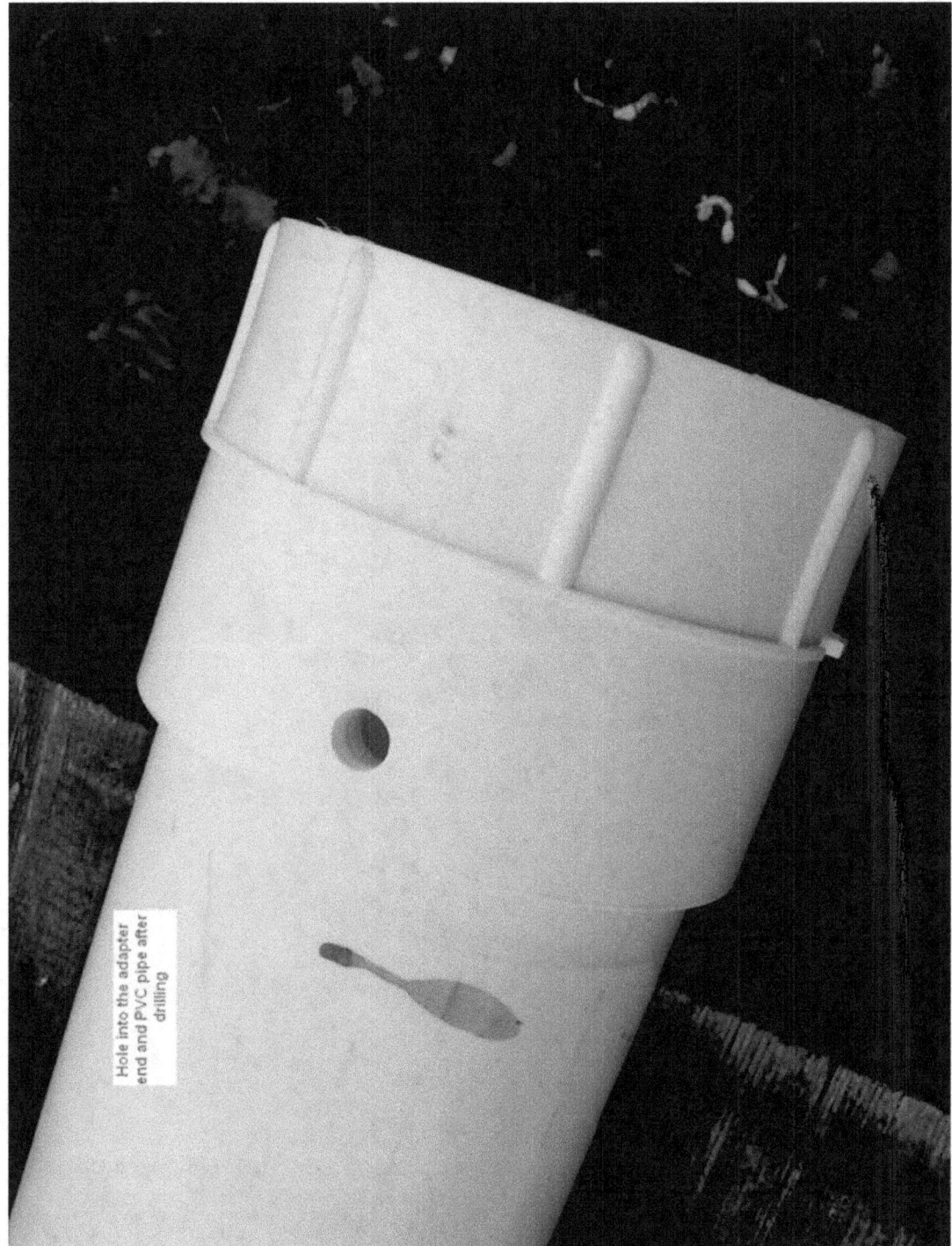

Hole into the adapter end and PVC pipe after drilling

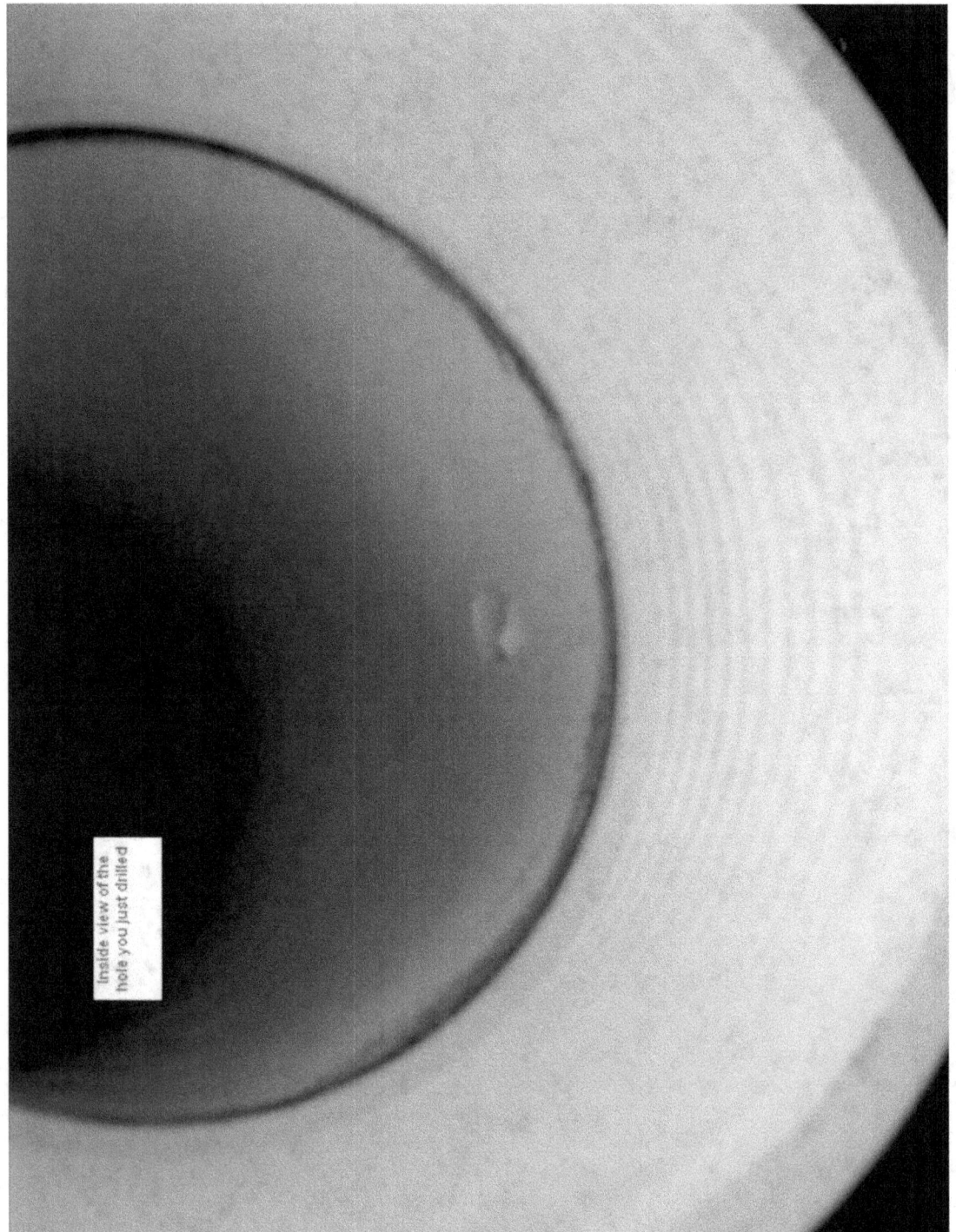

Inside view of the hole you just drilled

Using a ¾-18 tap, screw the tap into the hole that you just drilled until almost all the threads are in

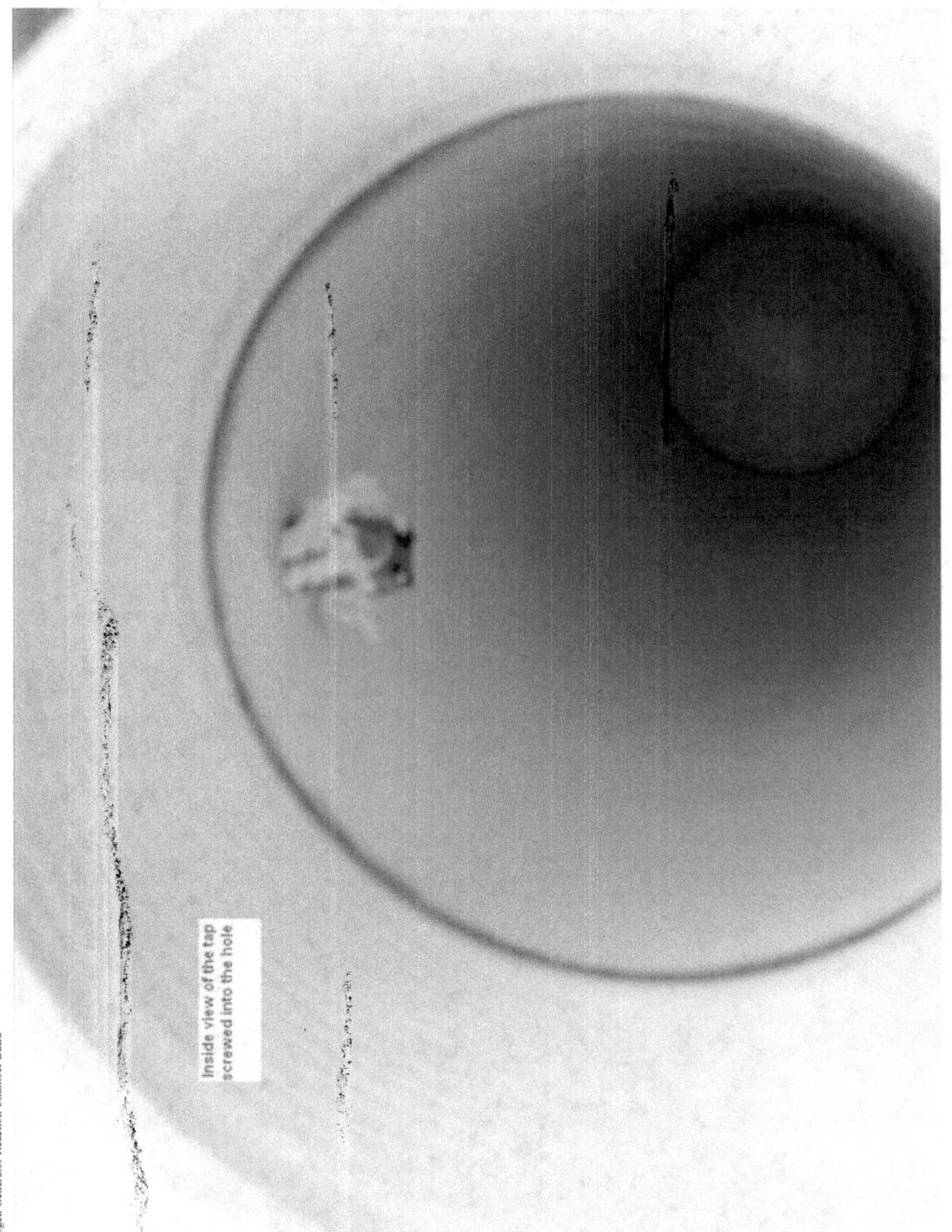

Inside view of the tap
screwed into the hole

This is the finished threaded hole after you remove the tap and clean the area of PVC debris

Repeat the last procedure with the center of the screw end cap

Hydrogen Generator Reaction Chamber Build

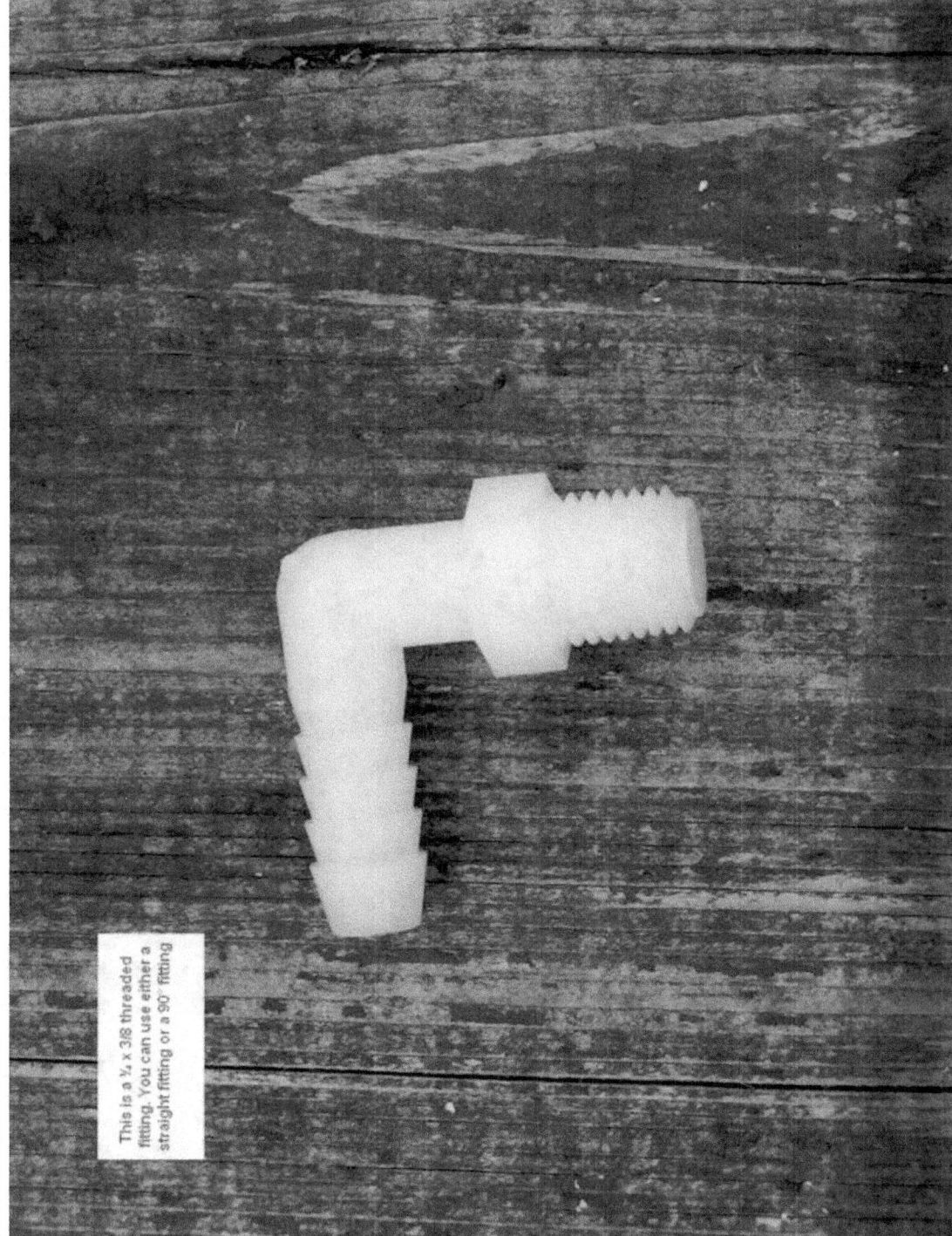

This is a ½ x 3/8 threaded fitting. You can use either a straight fitting or a 90° fitting

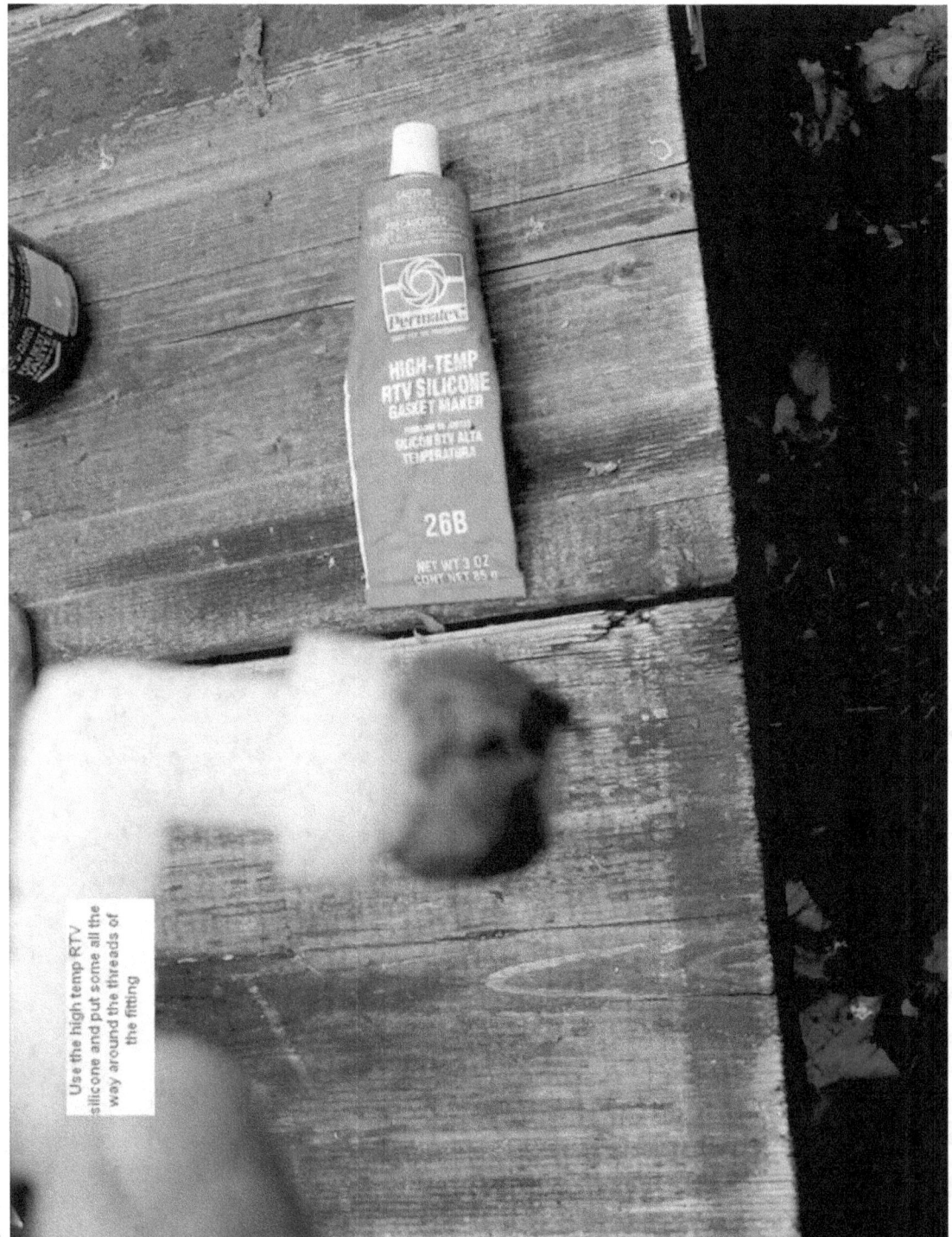

Use the high temp RTV silicone and put some all the way around the threads of the fitting

Screw the fitting into the threaded end cap until tight, and then smear the RTV around the base of the fitting to seal the bottom

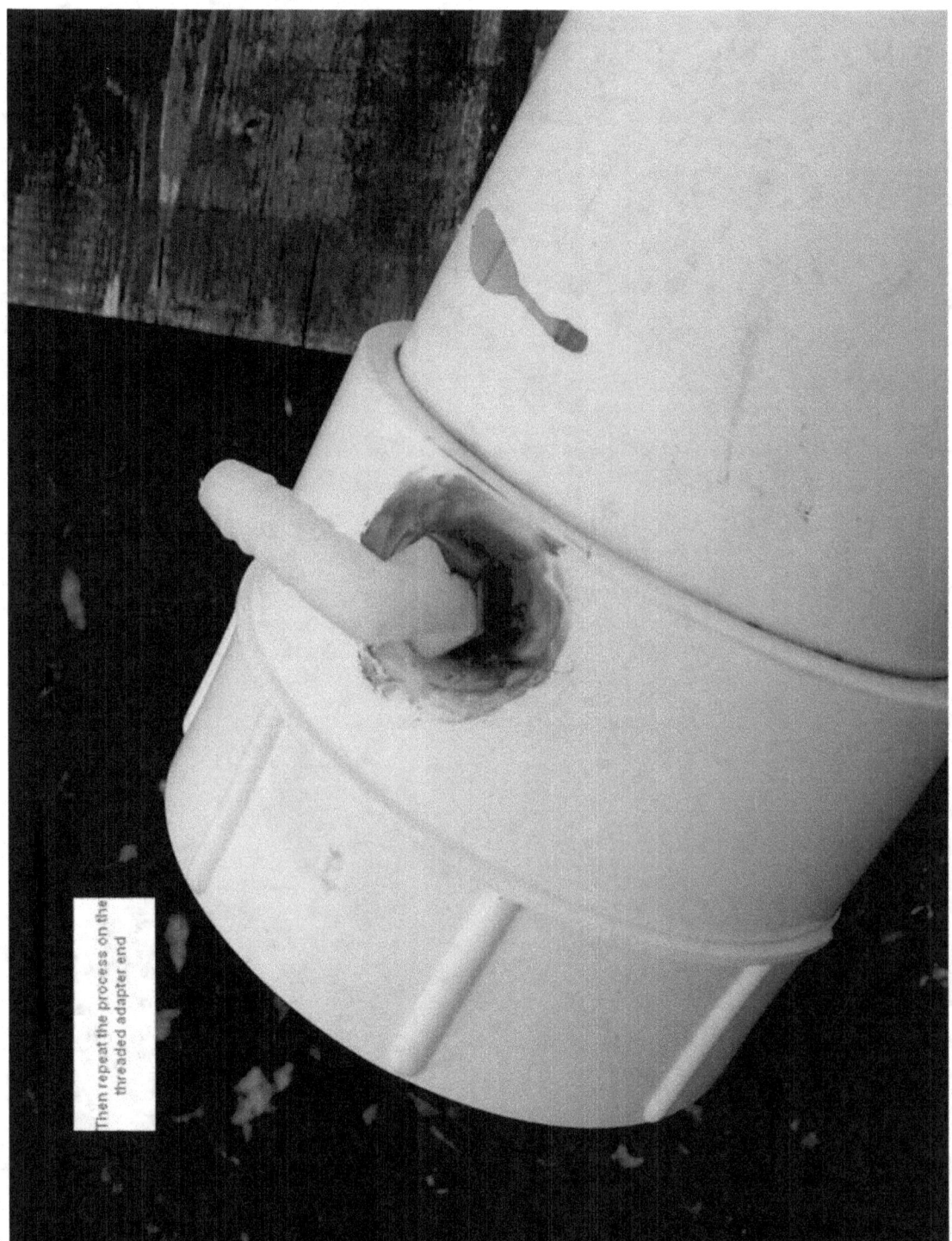

Then repeat the process on the threaded adapter end

Set the reaction chamber up on the adapter end. Using the RTV, seal the gap between the adapter end and the PVC pipe

#67L 03/19/

Flip the reaction chamber over and repeat the process on the solid end cap

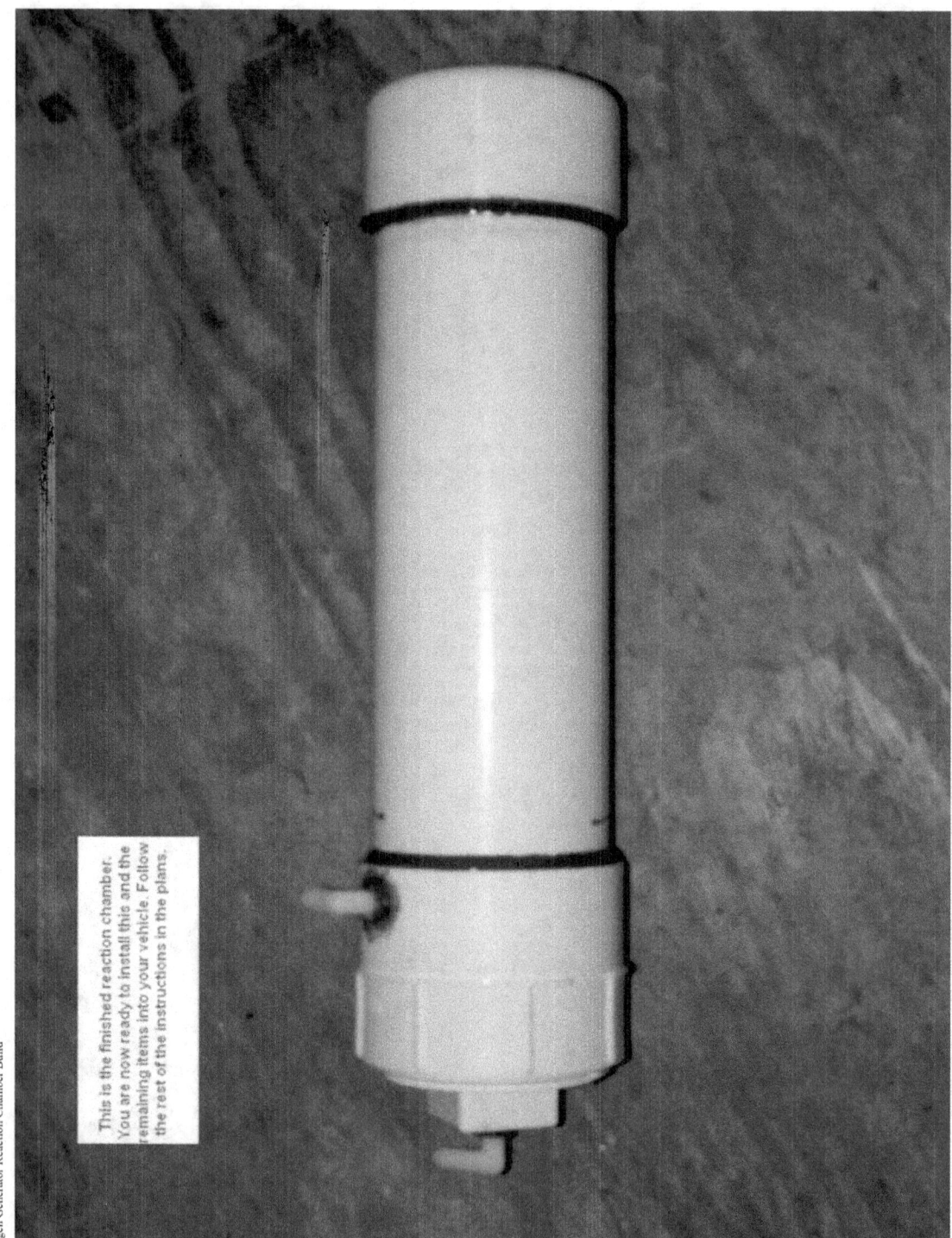

This is the finished reaction chamber. You are now ready to install this and the remaining items into your vehicle. Follow the rest of the instructions in the plans.